MY Learning Stations

TEACHER GUIDE
DIFFERENTIATED MATH INSTRUCTION

Grade 2

Bothell, WA • Chicago, IL • Columbus, OH • New York, NY

Copyright © 2013 The McGraw-Hill Companies, Inc.

All rights reserved. No part of this publication may be reproduced or distributed in any form or by any means, or stored in a database or retrieval system, without the prior written consent of The McGraw-Hill Companies, Inc., including, but not limited to, network storage or transmission, or broadcast for distance learning.

Permission is granted to reproduce the material contained on pages 31–60 on the condition that such material be reproduced only for classroom use; be provided to students, teachers, or families without charge; and be used solely in conjunction with *My Math*.

Send all inquiries to:
McGraw-Hill Education
8787 Orion Place
Columbus, OH 43240

ISBN: 978-0-02-117179-8
MHID: 0-02-117179-3

Printed in the United States of America.

7 8 9 10 RHR 18 17 16 15 14

Common Core State Standards© Copyright 2010. National Governors Association Center for Best Practices and Council of Chief State School Officers. All rights reserved.

Our mission is to provide educational resources that enable students to become the problem solvers of the 21st century and inspire them to explore careers within Science, Technology, Engineering, and Mathematics (STEM) related fields.

Contents

Components of My Learning Stations ii

Managing My Learning Stations iv

Activities in My Learning Stations viii

CHAPTER 1   Apply Addition and Subtraction Concepts 1

CHAPTER 2   Number Patterns. 3

CHAPTER 3   Add Two-Digit Numbers 5

CHAPTER 4   Subtract Two-Digit Numbers 6

CHAPTER 5   Place Value to 1,000 8

CHAPTER 6   Add Three-Digit Numbers 9

CHAPTER 7   Subtract Three-Digit Numbers 11

CHAPTER 8   Money. 12

CHAPTER 9   Data Analysis 14

CHAPTER 10   Time. 16

CHAPTER 11   Customary and Metric Lengths. 18

CHAPTER 12   Geometric Shapes and Equal Shares 21

Answers 23

Blackline Masters 31

Components of
MY Learning Stations

Differentiated Learning Stations

My Learning Stations is a collection of activity cards, literature, games, graphic novels, and problem-solving cards that can be used with each chapter of the *My Math* student edition. Differentiated instruction strategies for each component are provided within this teacher guide.

 The activities in *My Learning Stations* align to the Common Core State Standards and support the Standards for Mathematical Practice.

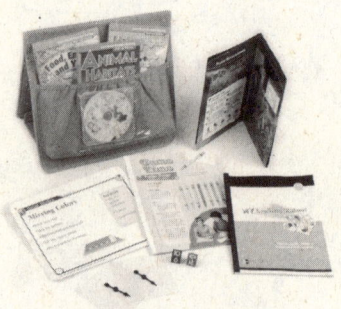

Carrying Case
Stores all of the components that make up the learning stations kit.

Learning Stations Teacher Guide
Includes instructions for using each component with approaching-level, on-level, and beyond-level students. Also includes answers and blackline masters.

KEY: Approaching Level   On Level   Beyond Level

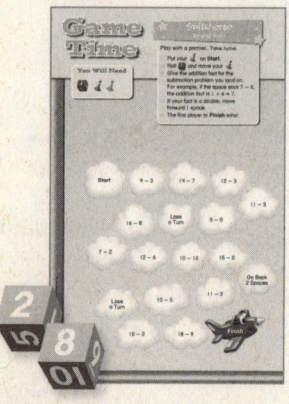

Games
Reinforce, challenge, and extend mathematical concepts being taught in the Student Edition. English is on one side and Spanish is on the other. Some of the materials needed to play the games are included in the kit.

* Provided in kit

 Graphic Novel

Introduce and revisit real-world mathematical situations. Animations bring the math content to life.

Real-World Problem Solving Readers (RWPS)

Fiction and non-fiction readers extend problem-solving skills and strategies and make real-world connections. Three leveled readers—approaching level, on level, and beyond level—are included in the kit.

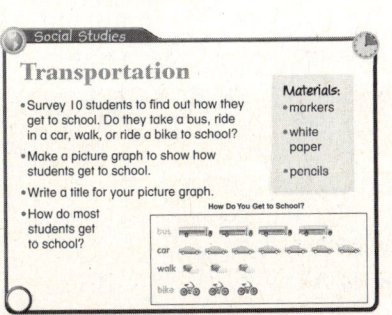

 Activity Cards

Give students an opportunity to learn mathematics through cross-curricular connections. English is on one side and Spanish is on the other.

 Problem Solving

Tie problem solving to other disciplines with real-world applications. English is on one side and Spanish is on the other. (Grades 3–5 only)

Managing MY Learning Stations

What are math learning stations?

Math learning stations are set up as places in the classroom where students work independently or in small groups exploring, practicing, or extending mathematical concepts. When working in math learning stations, students are engaged in problem-solving activities; they are reasoning; representing their thinking; communicating; and making connections between mathematical practices and content. *My Learning Stations* uses games, real-world problem solving readers, activity cards, and graphic novels to meet individual student needs. While students are involved in learning stations, the teacher works with individuals or meets with small groups in order to further differentiate math instruction.

How do I organize my classroom?

Evaluate the space in your classroom and think about how it can be arranged to best promote independent and small group instruction.

- Place the learning station easel in an area that is easily accessible to students.
- Make sure all supplies needed for each activity are available at the learning station easel.
- Ensure that the space you have chosen allows for easy cleanup.
- Encourage students to complete the activity at their desks, on the floor, on carpet tiles, at the computer, at an interactive whiteboard, at a bulletin board, or at tables. Allow students to spread out around the room to manage the level of noise and movement.
- Be sure the places where students choose to work are visible from where you may be working with a small group to allow you to monitor student engagement and progress.

> **What might this look like?**
> While you are working with a small group of eight students on the day's math lesson, two students are on the floor playing a game, four students are watching a graphic novel, and four students are at desks reading a real-world problem solving reader.

How do I get started?

During the first few weeks of school, develop and model procedures for using math learning stations.

- Discuss with the class what happens during math learning station time. Write student responses on a chart that you can display in the classroom. For example:

Math Learning Station Time

See	Hear	Think
students talking about math	math vocabulary being used	This is fun!
students taking turns	students explaining what they did	I can do math!
students sharing materials	students asking questions	I know how to do that!

- Math learning station time comes after a whole-class math lesson. This is not a time to introduce new concepts. It is a time for students to further explore, practice, or extend a previously taught math concept. To ensure students will be able to work independently during math learning station time, model what you expect students to do at each math learning station.

- For younger students and English language learners, create a brief list with simple drawings or pictures to help them remember what to do at each learning station.

- The time spent at a learning station will vary with the age of the students and the time of the school year. Kindergarten students may spend about 15 minutes in each learning station while first through fifth grade students may spend about 20 minutes.

- Going through all the learning stations in a chapter may take several days or weeks depending on the needs of the students in your class.

Managing
MY Learning Stations *continued*

- Create a system for managing any blackline masters needed at the learning stations.
  - Label a folder for each activity and put the blackline masters needed for that activity in the folder.

- Create a system for managing learning station work that is completed and ready for review. For example:
  - Label folders for each student or activity.
  - Label a box *Finished* in which students place completed work.
  - Use student logs or journals for recording.
  - Bring students back together in a large group and have them share any work product that was created at the learning station or tell what they did at the learning station.

- Decide on a rotation chart format to direct who goes to which learning station. If students are going to two learning stations, use a bell or timer to signal when to change learning stations.

What might a rotation chart look like?

Following is an example of one type of rotation chart you might use.

- For grades K-2, divide a piece of chart paper into five sections. Label the sections with the following titles: "Meet with Me"; Game; Graphic Novel; Real-World Problem Solving Reader; and Activity Card.

- For grades 3-5, divide a piece of chart paper into six sections. Label the sections with the following titles: "Meet with Me"; Game; Graphic Novel; Real-World Problem Solving Reader; Activity Card; and Problem-Solving Card.

- Write each student's name on a note card.

- Place student name cards in each learning station section on the chart to indicate which learning station they will be doing.

- If students are going to two learning stations, you will need to make a different color name card for each student and number the name cards 1 and 2. Then place the student name cards in each learning station section on the chart to indicate the order in which students do the stations.

How do I differentiate the learning stations?

Math learning stations allow you to differentiate for individual- and small-group needs of the students in the classroom.

- Determine station groupings

    - Use informal assessment (including observations) to determine individual student needs.

    - Decide whether groups will be homogeneous or heterogeneous.

    - Remember to keep reassessing and regrouping throughout the year.

    - Determine whether students will be working in a group or individually.

    - Use the buddy system to help with individualized work.

- After assessing individual student needs, refer to specific activity strategies in this Teacher Guide for differentiated instruction suggestions.

    AL identifies strategies for students who have needs that are below or approaching grade level.

    OL identifies strategies for students on grade level.

    BL identifies strategies for students who have needs that are above or beyond grade level.

- As students progress in concept understanding, they may repeat activities using different strategies, approaches, tools, and models.

> **What might this look like?**
> Strategies for playing a game may include having students who are *approaching level* work in teams of two to play, students who are *on level* play the game with the rules as written on the game board, and students who are *beyond level* bump the game up by using higher numbers.

Activities in MY Learning Stations

CHAPTER 1 **Apply Addition and Subtraction Concepts** .... **1**   CCSS   Use after Lesson

Game Switcheroo		2.OA.2	1-12
Graphic Novel Puppies!		2.OA.1	1-10
Real-World Problem Solving Reader How Many Seeds?		2.OA.1	1-9
Activity Card Letter Subtraction		2.OA.2	1-8

CHAPTER 2 **Number Patterns** .... **3**   CCSS   Use after Lesson

Game Count the Stars		2.NBT.2	2-2
Graphic Novel Flamingo Fun		2.NBT.2	2-4
Real-World Problem Solving Reader Geese on the Go		2.OA.4	2-5
Activity Card High-Frequency Words		2.OA.3	2-6

CHAPTER 3 **Add Two-Digit Numbers** .... **5**   CCSS   Use after Lesson

Game Pick Your Path		2.NBT.5	3-4
Graphic Novel Tickets to the Puppet Show		2.NBT.5	3-4
Activity Card Telephone Words		2.NBT.5	3-5

CHAPTER 4 **Subtract Two-Digit Numbers** .... **6**   CCSS   Use after Lesson

Game Hit the Target		2.NBT.5	4-5
Graphic Novel Swimming Fun		2.NBT.5	4-5
Real-World Problem Solving Reader Baseball's Hero		2.NBT.5	4-5
Activity Card Clay Sentences		2.NBT.5	4-5

CHAPTER 5	Place Value to 1,000	8	CCSS	Use after Lesson
	Game Butterfly Fun		2.NBT.3	5-3
	Graphic Novel Home Run Records		2.NBT.4	5-7
	Activity Card Comparing Stories		2.NBT.3	5-5

CHAPTER 6	Add Three-Digit Numbers	9	CCSS	Use after Lesson
	Game Three-Digit Fruit		2.NBT.7	6-6
	Graphic Novel Enchanted Palace Park		2.NBT.7	6-6
	Real-World Problem Solving Reader Lady Liberty		2.NBT.7	6-6
	Real-World Problem Solving Reader Moving Along		2.NBT.7	6-6
	Activity Card How Many Calories?		2.NBT.7	6-6

CHAPTER 7	Subtract Three-Digit Numbers	11	CCSS	Use after Lesson
	Game Subtract It!		2.NBT.7	7-6
	Graphic Novel Ice Cream Favorites		2.NBT.7	7-6
	Activity Card The Boiling Point		2.NBT.7	7-6

CHAPTER 8	Money	12	CCSS	Use after Lesson
	Game Money "Cents"		2.MD.8	8-5
	Graphic Novel Pinwheel, Please!		2.MD.8	8-5
	Real-World Problem Solving Reader Our Grandma's Life		2.MD.8	8-5
	Activity Card Money Words		2.MD.8	8-5

CHAPTER 9	Data Analysis	14	CCSS	Use after Lesson
	Game Spring Sports		2.MD.10	9-5
	Graphic Novel And the Winner Is…		2.MD.10	9-3
	Activity Card Say It With Pictures		2.MD.10	9-3

Activities in MY Learning Stations (continued)

CHAPTER 10 Time 16

		CCSS	Use after Lesson
Game Time to Feast		2.MD.7	10-5
Graphic Novel Pizza Time		2.MD.7	10-5
Real-World Problem Solving Reader A Mountain of Presidents		2.MD.7	10-6
Activity Card Time Journal		2.MD.7	10-6

CHAPTER 11 Customary and Metric Lengths 18

		CCSS	Use after Lesson
Game Inching Along		2.MD.1	11-1
Graphic Novel Surfboard Selector		2.MD.4	11-4
Real-World Problem Solving Reader A Magnet's Strength		2.MD.4	11-9
Real-World Problem Solving Reader Fossils Over Time		2.MD.4	11-9
Real-World Problem Solving Reader Animals Big and Small		2.MD.4	11-10
Activity Card Centimeter Caterpillar		2.MD.3	11-7

CHAPTER 12 Geometric Shapes and Equal Shares 21

		CCSS	Use after Lesson
Game The Equalizer		2.G.3	12-6
Graphic Novel Soccer Shape-Up		2.G.1	12-5
Graphic Novel The Case of the Missing Cake		2.G.3	12-7
Real-World Problem Solving Reader Homes of All Shapes		2.G.1	12-5

Answers 23
Blackline Masters 31

CHAPTER 1 Apply Addition and Subtraction Concepts

The following learning station activities provide differentiated instructional strategies for helping students fluently add and subtract within 20 and relate addition and subtraction.

 2.OA.2   Use after Lesson 1-12

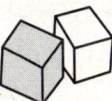

Game Switcheroo

Materials two different color game pieces, 0–5 number cube*, blackline master page 31 or blackline master page 32 are needed to extend the game

- **AL** Have students play the game, giving the answers to the subtraction problems rather than the related addition fact.
- **OL** Have students play the game with the rules as written.
- **BL** Have students give the entire fact family for each fact that is landed on.

Extend the Game Have students use blackline master page 31 or blackline master page 32 to create an alternate game board using addition facts. Differentiate the new game by adjusting and following the suggestions above.

 2.OA.1   Use after Lesson 1-10

Graphic Novel Puppies!

Materials blackline master pages 33 and 34, crayons, pencil

Watch the graphic novel as a class and discuss the following questions.

- What is happening? *Sample answer: The children are looking at puppies at the pet store.*
- How did they find how many puppies there are in all? *Sample answer: They added the number of puppies in each bed.*
- What do you think will happen to the puppies in the pet store? *Sample answer: They will be sold, and there will be fewer of them.*

- **AL** Have students solve the graphic novel problem using blackline master page 33.
- **OL** Have students solve the graphic novel problem using blackline master page 33. On the back of the blackline master, have students extend the graphic novel by writing another subtraction problem that involves the puppies at the pet store. Have students ask a friend to solve the new problem.
- **BL** Have students solve the graphic novel problem using blackline master page 33. Have students use blackline master page 34 to write their own graphic novels with a math problem that uses subtraction to solve. Let students share their new graphic novels with a friend and ask the friend to solve the problem.

KEY: 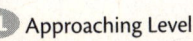 Approaching Level    On Level   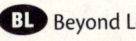 Beyond Level   * Provided in kit

CHAPTER 1   Apply Addition and Subtraction Concepts   (continued)

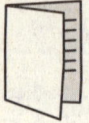

 2.OA.1   Use after Lesson 1-9

Real-World Problem Solving Reader *How Many Seeds?*

Summary In *How Many Seeds?*, students study a diagram to follow the life cycle of a plant. Students also investigate the numbers of seeds in a variety of plants and read charts to compare and analyze data.

Materials paper, pencil

  Have students read the book and answer Exercises 1, 3, 4, and 6 with a partner.

  Have students read the book and answer Exercises 1, 3, 4, and 6.

  Have students read the book and answer Exercises 1, 3, 4, 5, and 6.

2.OA.2   Use after Lesson 1-8

Activity Card Letter Subtraction

Materials paper strips, scrap paper, pencil

  Give students strips with words already written on them. Be sure no word is longer than 12 letters.

  Have students complete the Activity Card as written.

  Have students use specific vocabulary terms for their word choices—such as all animal names, all math vocabulary, or all people's names. Ask students to use words with no more than 18 letters.

CHAPTER 2 Number Patterns

The following learning station activities provide differentiated instructional strategies for helping students skip count and use addition and subtraction to solve word problems.

 2.NBT.2   Use after Lesson 2-2

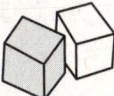

Game Count the Stars

Materials two different color connecting cubes, number line, 0–5 number cube*, 5–10 number cube*

- **AL** Have students make use of a number line to help them count by 5s.
- **OL** Have students play the game with the rules as written.
- **BL** Have students play the game skip counting by 2s.

 2.NBT.2   Use after Lesson 2-4

Graphic Novel Flamingo Fun

Materials blackline master pages 35 and 36, crayons, pencil

Watch the graphic novel as a class and discuss the following questions.

- What is happening? *Sample answer: The children are looking at the flamingos at Ocean World. The boy is trying to count the number of flamingo legs he sees.*

- What is the problem? *Sample answer: The girls' talking is distracting the boy and it is taking him too long to count all the flamingo legs. He needs to find a quicker way to count them.*

- **AL** Have students solve the graphic novel problem using blackline master page 35
- **OL** Have students solve the graphic novel problem using blackline master page 35. On the back of the blackline master, have students extend the graphic novel by drawing another cell to show what could happen next.
- **BL** Have students solve the graphic novel problem using blackline master page 35. Have students use blackline master page 36 to write their own graphic novels with a problem that uses skip counting to solve. Let students share their new graphic novels with a friend and ask the friend to solve the problem.

CHAPTER 2 **Number Patterns** (continued)

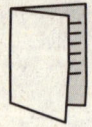

CCSS 2.OA.4   Use after Lesson 2-5

Real-World Problem Solving Reader *Geese on the Go*

Summary *Geese on the Go* overviews the migration of Canada geese. It includes migration routes, distances, and durations. The book also includes information about Canada geese families. Students will use addition and subtraction to answer questions about Canada geese.

Materials 16 counters, paper, pencil

- **AL** Have students work in pairs. Have student pairs read the book and use counters to answer Exercises 1, 3, and 5.
- **OL** Have students read the book and answer Exercises 1, 2, 3, and 5.
- **BL** Have students read the book and answer Exercises 1–5.

CCSS 2.OA.3   Use after Lesson 2-6

Activity Card High-Frequency Words

Materials strips of paper with high-frequency words written on them

- **AL** Write words on the strips with twelve or fewer letters.
- **OL** Have students complete the Activity Card as written.
- **BL** Have students write their own words on strips of paper. Have students ask a friend to choose a word, count the number of letters, and tell whether there is an even or odd number of letters.

CHAPTER 3 Add Two-Digit Numbers

The following learning station activities provide differentiated instructional strategies for helping students add two-digit numbers using strategies based on place value and properties of operations.

 2.NBT.5   Use after Lesson 3-4

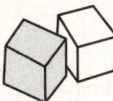

Game Pick Your Path

Materials two different color game pieces, 0–5 number cube*, base-ten blocks, paper, pencil

- **AL** Have students use base-ten blocks to help find the sums.
- **OL** Have students play the game with the rules as written.
- **BL** Have students do all of the addition mentally.

Extend the Game Have students play the game without using any spaces that the other player has already landed on. Differentiate the new game by following the suggestions above.

 2.NBT.5   Use after Lesson 3-4

Graphic Novel Tickets to the Puppet Show

Materials blackline master pages 37 and 38, crayons, pencil

Watch the graphic novel as a class and discuss the following questions.

- What is happening? *Sample answer: The children are selling tickets to a puppet show. Their goal is to sell 90 tickets.*
- What is the problem? *Sample answer: The children need to figure out if they met their goal of selling 90 tickets.*

- **AL** Have students solve the graphic novel problem using blackline master page 37.
- **OL** Have students solve the graphic novel problem using blackline master page 37. On the back of the blackline master, have students extend the graphic novel by writing another two-digit addition problem. Have students ask a friend to solve the new problem.
- **BL** Have students solve the graphic novel problem using blackline master page 37. Have students use blackline master page 38 to write their own graphic novels with a problem that uses two-digit addition to solve. Let students share their new graphic novels with a friend and ask the friend to solve the problem.

 2.NBT.5   Use after Lesson 3-5

Activity Card Telephone Words

Materials telephone keypad (with numbers and letters) drawn on construction paper, paper with the words *it*, *are*, *best*, *sun*, and *friend* written on it

- **AL** Have students choose a word from the list. Have them match the letters with numbers from the keypad and add the numbers to find the sum for the telephone word. Repeat until all the words on the list have been used.
- **OL** Have students complete the Activity Card as written.
- **BL** Have students work in pairs and write their own words on paper and then follow the directions on the Activity Card.

5

CHAPTER 4   Subtract Two-Digit Numbers

The following learning station activities provide differentiated instructional strategies for helping students subtract two-digit numbers using strategies based on place value and properties of operations.

 2.NBT.5   Use after Lesson 4-5

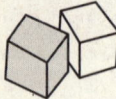

Game Hit the Target

Materials two different color game pieces, 5–10 number cube*, 10 two-colored counters, hundred chart or base-ten blocks

- **AL** Have students play the game using a hundred chart or base-ten blocks to help them find the correct answer.
- **OL** Have students play the game with the rules as written.
- **BL** Have students give both subtraction and addition problems that will result in the desired answer.

Extend the Game Have students create an alternate game board using different numbers, which will require different problems. Differentiate the new game by following the suggestions above.

 2.NBT.5   Use after Lesson 4-5

Graphic Novel Swimming Fun

Materials blackline master pages 39 and 40, crayons, pencil

Watch the graphic novel as a class and discuss the following questions.

- What is happening? *Sample answer: The children are going to take turns diving for 20 rings. One of the girls thinks it will be easy to get all 20 rings and begins diving for them.*

- What is the problem? *Sample answer: The girl is tired and surprised to find that she doesn't have all 20 rings. The children will have to find how many rings she has so far and then determine how many more she has to get.*

- **AL** Have students solve the graphic novel problem using blackline master page 39.
- **OL** Have students solve the graphic novel problem using blackline master page 39. On the back of the blackline master, have students change the number of rings Julie brings up each time, and ask a friend to find how many more times she has to dive.
- **BL** Have students solve the graphic novel problem using blackline master page 39. Have students use blackline master page 40 to write their own graphic novels with a problem that uses two-digit subtraction to solve. Let students share their new graphic novels with a friend and ask the friend to solve the problem.

6

CHAPTER 4   **Subtract Two-Digit Numbers** *(continued)*

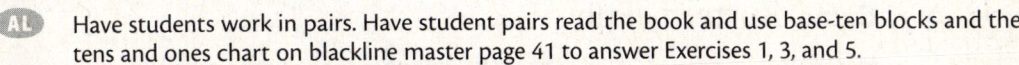

2.NBT.5   Use after Lesson 4-5

Real-World Problem Solving Reader *Baseball's Hero*

Summary In *Baseball's Hero*, students will read about Jackie Robinson's career in Major League Baseball. They will use two-digit addition and two-digit subtraction to answer questions about baseball and Mr. Robinson's accomplishments.

Materials base-ten blocks, blackline master page 41, paper, pencil

- **AL** Have students work in pairs. Have student pairs read the book and use base-ten blocks and the tens and ones chart on blackline master page 41 to answer Exercises 1, 3, and 5.
- **OL** Have students read the book and answer Exercises 1, 3, 4, and 5.
- **BL** Have students read the book and answer Exercises 1–5.

 2.NBT.5   Use after Lesson 4-5

Activity Card *Clay Sentences*

Materials modeling clay; tens and ones chart on blackline master page 41; base-ten blocks; connecting cubes, or a whiteboard; index cards with two-digit subtraction problems

- **AL** Provide index cards with two-digit subtraction problems that student pairs will model.
- **OL** Have students complete the Activity Card as written.
- **BL** Have students model number sentences that subtract a two-digit number from a two-digit number.

CHAPTER 5 Place Value to 1,000

The following learning station activities provide differentiated instructional strategies for helping students read and write numbers to 1,000 using expanded form and compare three-digit numbers.

 2.NBT.3   Use after Lesson 5-3

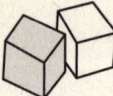

Game Butterfly Fun

Materials two different color crayons, 0–5 number cube*, paper, pencil

- **AL** Have students roll the number cube two times to make and read a two-digit number.
- **OL** Have students play the game with the rules as written.
- **BL** Have students roll the number cube four times to make and read a four-digit number.

 2.NBT.4   Use after Lesson 5-7

Graphic Novel Home Run Records

Materials blackline master pages 42 and 43, crayons, pencil

Watch the graphic novel as a class and discuss the following questions.

- What is happening? *Sample answer: The children are comparing home run records of their favorite baseball players.*
- What is the problem? *Sample answer: They want to know which player has hit the most home runs.*

- **AL** Have students solve the graphic novel problem using blackline master page 42.
- **OL** Have students solve the graphic novel problem using blackline master page 42. On the back of the blackline master, have students extend the graphic novel by writing another problem using new home run record numbers. Have students ask a friend to solve the new problem.
- **BL** Have students solve the graphic novel problem using blackline master page 42. Have students use blackline master page 43 to write their own graphic novels with a problem that involves comparing three-digit numbers. Let students share their new graphic novels with a friend and ask the friend to solve the problem.

 2.NBT.3   Use after Lesson 5-5

Activity Card Comparing Stories

Materials 10 index cards with the numbers 10, 100, 1,000, 500, 437, 621, 78, 981, 284, and 588 written on them, pencil, paper

- **AL** Have students work in pairs and use the cards with 10, 78, 100, and 500 written on them to complete the activity.
- **OL** Have students complete the Activity Card as written.
- **BL** Have students complete the Activity Card as written, using their own numbers in their tall tales instead of choosing two number cards.

8

CHAPTER 6 Add Three-Digit Numbers

The following learning station activities provide differentiated instructional strategies for helping students add within 1,000.

 2.NBT.7   Use after Lesson 6-6

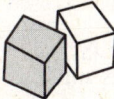
Game Three-Digit Fruit

Materials 10 two-color counters, tens and ones chart on blackline master page 41, base-ten blocks, pencil, paper

  Have students use the tens and ones chart on blackline master page 41 and base-ten blocks to solve the problems.

  Have students play the game with the rules as written.

  Have students write their own addition problems that involve regrouping. Tape the new problems on top of the old problems, and have them play the game.

 2.NBT.7   Use after Lesson 6-6

Graphic Novel Enchanted Palace Park

Materials blackline master pages 44 and 45, crayons, pencil

Watch the graphic novel as a class and discuss the following questions.

- What is happening? *Sample answer: The children are talking about getting passes to the Enchanted Palace Park.*

- What is the problem? *Sample answer: They want to find out how much it will cost for annual passes for Julie and her brother.*

  Have students solve the graphic novel problem using blackline master page 44.

  Have students solve the graphic novel problem using blackline master page 44. On the back of the blackline master, have students extend the graphic novel by finding out how much it will cost for annual passes for Julie and her nine-year old sister.

  Have students solve the graphic novel problem using blackline master page 44. Have students use blackline master page 45 to write their own graphic novels with a problem that uses three-digit addition to solve. Let students share their new graphic novels with a friend and ask the friend to solve the problem.

CHAPTER 6 **Add Three-Digit Numbers** (continued)

  2.NBT.7   Use after Lesson 6-6

Real-World Problem Solving Reader *Lady Liberty*

Summary In *Lady Liberty*, students explore the Statue of Liberty. They will discover her history and use her dimensions to solve problems. They will also use their mapping skills to explore the area near the statue.

Materials base-ten blocks, blackline master page 41, paper, pencil

- **AL** Have students read the book with a partner. Have student partners use base-ten blocks and the tens and ones chart on blackline master page 41 to answer Exercise 1.
- **OL** Have students read the book and answer Exercise 1. Then have them look at page 16 and find how many steps there are altogether.
- **BL** Have students read the book and answer Exercise 1. Have students look at page 12 and find out how much it would cost two people to visit the Statue of Liberty in June. Then have them look at page 16 and find how many steps there are altogether.

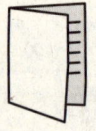

  2.NBT.7   Use after Lesson 6-6

Real-World Problem Solving Reader *Moving Along*

Summary In *Moving Along*, students follow the experiences of a family moving from Oklahoma to California in 1950 along Route 66. The book provides experiences with time, distance, and reading maps.

Materials base-ten blocks, blackline master page 41, paper, pencil

- **AL** Have students read the book with a partner. Have student partners use base-ten blocks and the tens and ones chart on blackline master page 41 to answer Exercise 3.
- **OL** Have students read the book and answer Exercise 3. Then have students reread the postcard on page 7 and find out how far the family will be from Hydro by sundown.
- **BL** Have students read the book and answer Exercise 3. Have them reread the postcard on page 7 and find out how far the family will be from Hydro by sundown. Then have them reread the postcard on page 13 and find how old some healthy Ponderosa pines live to be if they live 77 years longer.

2.NBT.7   Use after Lesson 6-6

Activity Card How Many Calories?

Materials food labels, base-ten blocks, blackline master page 41, paper, pencil

- **AL** Have students use base-ten blocks and the tens and ones chart on page 41 to help them add the number of calories on the two food labels.
- **OL** Have students complete the Activity Card as written.
- **BL** Have students complete the Activity Card as written. Then have students repeat the activity selecting food labels that will result in the lowest calorie count and the highest calorie count.

CHAPTER 7 Subtract Three-Digit Numbers

The following learning station activities provide differentiated instructional strategies for helping students subtract within 1,000.

CCSS 2.NBT.7   Use after Lesson 7-6

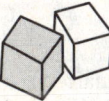

Game Subtract It!

Materials two different color game pieces, 0–5 number cube*, 12 two-color counters, tens and ones chart on blackline master page 41, base-ten blocks

- **AL** Have students use the tens and ones chart on blackline master page 41 and base-ten blocks to solve the problems.
- **OL** Have students play the game with the rules as written.
- **BL** Have students name another subtraction problem that would give the same difference before they move their game pieces.

 2.NBT.7   Use after Lesson 7-6

Graphic Novel Ice Cream Favorites

Materials blackline master pages 46 and 47, crayons, pencil

Watch the graphic novel as a class and discuss the following questions.

- What is happening? *Sample answer: The children took a vote to find the favorite flavor of ice cream. They have just counted all the votes. Chocolate ice cream won more votes than strawberry ice cream.*
- What is the problem? *Sample answer: There is a disagreement over whether chocolate ice cream won by a lot of votes or if it just barely won. They have to find out how many more people chose chocolate ice cream over strawberry ice cream as their favorite flavor.*

- **AL** Have students solve the graphic novel problem using blackline master page 46.
- **OL** Have students solve the graphic novel problem using blackline master page 46. On the back of the blackline master, have students extend the graphic novel by drawing another cell to show what could happen next.
- **BL** Have students solve the graphic novel problem using blackline master page 46. Have students use blackline master page 47 to write their own graphic novels with a problem that uses three-digit subtraction to solve. Let students share their new graphic novels with a friend and ask the friend to solve the problem.

 2.NBT.7   Use after Lesson 7-6

Activity Card The Boiling Point

Materials basket, slips of paper with three-digit numbers from 100 to 211 written on them, tens and ones chart on blackline master page 41, base-ten blocks, paper, pencil

- **AL** Have students use the tens and ones chart on blackline master page 41 and base-ten blocks to help them subtract.
- **OL** Have students complete the Activity Card as written.
- **BL** Have students complete the Activity Card as written. Then have students select another slip of paper and write a word problem using the boiling point of water and the number on the slip of paper.

11

CHAPTER 8 Money

The following learning station activities provide differentiated instructional strategies for helping students identify and count pennies, nickels, dimes, and quarters and subtract two-digit numbers.

 2.MD.8   Use after Lesson 8-5

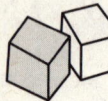

Game Money "Cents"

Materials two different color game pieces, 0–5 number cube*, coins (pennies, nickels, dimes, quarters), paper, pencil

- **AL** As students play the game, have them work together to model the money amounts.
- **OL** Have students play the game with the rules as written.
- **BL** Have students exchange for coins with greater values when possible.

 2.MD.8   Use after Lesson 8-5

Graphic Novel Pinwheel, Please!

Materials blackline master pages 48 and 49, dimes, nickels, crayons, pencil

Watch the graphic novel as a class and discuss the following questions.

- What is happening? *Sample answer: The children are in The Party Store. One of the girls is looking for a pinwheel to buy for her sister's birthday.*
- What is the problem? *Sample answer: They have to find out if she has enough money to buy the red pinwheel.*

- **AL** Have students use coins to solve the graphic novel problem on blackline master page 48.
- **OL** Have students solve the graphic novel problem using blackline master page 48. On the back of the blackline master, have students change the coins Sophia has and ask a friend to determine whether Sophia has enough money to get her sister a pinwheel.
- **BL** Have students solve the graphic novel problem using blackline master page 48. Have students use blackline master page 49 to write their own graphic novels with a problem that requires students to determine whether someone has enough money to buy an item. Let students share their new graphic novels with a friend and ask the friend to solve the problem.

CHAPTER 8 Money (continued)

2.MD.8 Use after Lesson 8-5

Real-World Problem Solving Reader *Our Grandma's Life*

Summary In *Our Grandma's Life*, a young girl talks with her grandmother about the grandmother's life in the early 1960s. They compare life at that time with life now. Students will also use map skills and math skills to understand comparisons and measurements.

Materials coins—three dimes, five nickels, 20 pennies, paper, pencil

- **AL** Have students read the book with a partner and answer Exercises 3 and 4.
- **OL** Have students read the book and answer Exercises 1, 2, and 4.
- **BL** Have students read the book and answer Exercises 1–4.

 2.MD.8 Use after Lesson 8-5

Activity Card Money Words

Materials vocabulary cards with the words *penny, nickel, dime, quarter, dollar,* and *dollar sign* written on them; crayons; paper; pencil

- **AL** Have students choose one card from the vocabulary cards *penny, nickel, dime, quarter,* and *dollar* and write the following sentence model: A _____ is equal to _____. on their papers. Students should complete the sentence with the coin and its value.
- **OL** Have students complete the Activity Card as written.
- **BL** Have students complete the Activity Card as written, writing two sentences using the vocabulary word they chose in each sentence.

13

CHAPTER 9 Data Analysis

The following learning station activities provide differentiated instructional strategies for helping students solve problems using information presented on picture graphs and bar graphs.

 2.MD.10   Use after Lesson 9-5

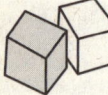
Game Spring Sports

Materials three different color crayons, transparent spinner* (assign each player a color on the color wheel), tally chart (for recording results of spins), bar graph outline (matching the bar graph on the game board), 0–5 number cube*, paper, pencil

- **AL** Have students record the result of each spin in a tally chart. The first player whose color accumulates six tally marks wins the game.
- **OL** Have students play the game with the rules as written.
- **BL** Have students select a number on a 0-5 number cube and a color on the color wheel. Players take turns tossing the cube and spinning the arrow at the same time. Players shade in a box on their bar graphs only when the cube lands on their number and the arrow lands on their color.

Extend the Game Have students play an opposite form of the game. The player whose color the spinner lands on is the only person who does not shade in a box of his or her bar graph. Players who reach the finish line are out of the race. The last person to reach the finish line is the winner. Differentiate the new game by following and adjusting the suggestions above.

 2.MD.10   Use after Lesson 9-3

Graphic Novel And the Winner Is…

Materials blackline master pages 50 and 51, crayons, pencil

Watch the graphic novel as a class and discuss the following questions.

- What is happening? *Sample answer: The children are recording votes for a class field trip.*
- Where are the possible places they can visit for the field trip? *zoo, art museum, and park*
- What is the problem? *Sample answer: The children have to find out which place received the most votes.*

- **AL** Have students solve the graphic novel problem using blackline master page 50.
- **OL** Have students solve the graphic novel problem using blackline master page 50. On the back of the blackline master, have students extend the graphic novel by changing the data in the graph and writing a new question. Have students ask a friend to answer the new question.
- **BL** Have students solve the graphic novel problem using blackline master page 50. Have students use blackline master page 51 to write their own graphic novels with a problem that uses information from a graph to solve. Let students share their new graphic novels with a friend and ask the friend to solve the problem.

14

CHAPTER 9 Data Analysis *(continued)*

2.MD.10 Use after Lesson 9-3

Activity Card Say It With Pictures

Materials crayons, grid paper, birth months of students in the class tallied on a sheet of paper, paper, pencil

- **AL** Have students complete the Activity Card as written.
- **OL** Have students complete the Activity Card as written. Then have them write on another piece of paper the month that has the most birthdays and the month that has the least number of birthdays.
- **BL** Have students complete the Activity Card as written. Have them find the difference between the month that has the most birthdays and the month that has the least number of birthdays.

CHAPTER 10 Time

The following learning station activities provide differentiated instructional strategies for helping students write and tell time to the nearest hour and half hour using analog and digital clocks.

 2.MD.7   Use after Lesson 10-5

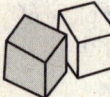

Game Time to Feast

Materials two different color crayons, blackline master page 52, 0–5 number cube*

  Have students play the game in teams of two.

  Have students play the game with the rules as written.

  Have students make a different game board with a different theme. For example, they might choose an airport theme, showing different times that planes arrive or leave.

 2.MD.7   Use after Lesson 10-5

Graphic Novel Pizza Time

Materials blackline master pages 53 and 54, crayons, pencil

Watch the graphic novel as a class and discuss the following questions.

- What is happening? *Sample answer: The girls are eating pizza before Kaya goes to gymnastic practice.*
- What does Kaya's mother remind them that they need to do? *Sample answer: They need to leave for gymnastics in 40 minutes.*
- What do they need to do before Kaya goes to gymnastics? *Sample answer: They need to take Sophia home.*

  Have students solve the graphic novel problem using blackline master page 53.

  Have students solve the graphic novel problem using blackline master page 53. On the back of the blackline master, have students extend the graphic novel by writing another problem that changes the amount of time Kaya has to get to gymnastics class. Have students ask a friend to solve the new problem.

  Have students solve the graphic novel problem using blackline master page 53. Have students use blackline master page 54 to write their own graphic novels with a problem that changes the time the girls leave to take Sophia home, and the time they have to get to gymnastics class. Let students share their new graphic novels with a friend and ask the friend to solve the problem.

16

CHAPTER 10 Time (continued)

 2.MD.7 Use after Lesson 10-6

Real-World Problem Solving Reader *A Mountain of Presidents*

Summary In *A Mountain of Presidents*, students see the process of carving Mt. Rushmore. In addition to using measuring and mapping skills, they will be able to use photographs to develop an understanding of scale and structure.

Materials blackline master page 52, paper, pencil

- **AL** Have students read the book with a partner. Have student pairs use the clock on blackline page 52 to answer Exercises 2 and 3.
- **OL** Have students read the book and answer Exercises 2–4.
- **BL** Have students read the book and answer Exercises 2–4. Have students look at page 13. In the summer, the park closes at 10:00 P.M. In the winter, the park closes 5 hours earlier. Have students find what time the park closes in the winter.

 2.MD.7 Use after Lesson 10-6

Activity Card Time Journal

Materials two sheets of paper folded and stapled into a six-page booklet; four sheets of paper folded and stapled into a ten-page booklet; pencil, crayons, or markers

- **AL** Have students complete the Activity Card as written, writing only one word to tell what happens at each of the times.
- **OL** Have students complete the Activity Card as written, writing one sentence on each page telling what happens during that time.
- **BL** Have students complete the Activity Card as written, using four sheets of paper folded and stapled into a 14-page booklet. Have students include the following times: 12 o'clock, half past 12, 1:00, 1:30, 2 o'clock, half past 2, 3 o'clock, and 3:30.

17

CHAPTER 11 Customary and Metric Lengths

The following learning station activities provide differentiated instructional strategies for helping students measure in inches, measure to determine how much longer or shorter one object is than another, and estimate lengths using units of centimeters.

 2.MD.1   Use after Lesson 11-1

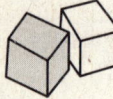

Game Inching Along

Materials two different color game pieces, 0–5 number cube*, blackline master page 31 or blackline master page 32

  Have students play the game as three-person teams, with each member performing one step of the turn.

  Have students play the game with the rules as written.

  Have students use blackline master page 31 or blackline master page 32 to create their own six-item chart with different objects.

Extend the Game Have students write the name of the object and its corresponding unit of measure. Then have them write the name of another object of similar size. Differentiate the new game by following the suggestions above.

 2.MD.4   Use after Lesson 11-4

Graphic Novel Surfboard Selector

Materials blackline master pages 55 and 56, crayons, pencil

Watch the graphic novel as a class and discuss the following questions.

- What is happening? *Sample answer: Kaya wants to go surfing and she is hoping to find a surfboard to rent.*

- What is the problem? *Sample answer: They have to figure out how tall Kaya is in inches and then determine how long the surfboard that she rents should be.*

  Have students solve the graphic novel problem using blackline master page 55.

  Have students solve the graphic novel problem using blackline master page 55. On the back of the blackline master, have students extend the graphic novel by drawing another cell to show what could happen next.

  Have students solve the graphic novel problem using blackline master page 55. Have students use blackline master page 56 to write their own graphic novels with a problem that changes the lengths of the surfboards and Kaya's height. Let students share their new graphic novels with a friend and ask the friend to read the graphic novel and solve the problem.

CHAPTER 11 **Customary and Metric Lengths** (continued)

CCSS 2.MD.4   Use after Lesson 11-9

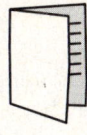

Real-World Problem Solving Reader *A Magnet's Strength*

Summary In *A Magnet's Strength*, students explore a variety of magnets and how we use them in everyday life. The book also presents experiments students can replicate and graphs to analyze and evaluate.

Materials paper, pencil

- **AL** Have students read the book with a partner and answer Exercise 4. Have students use the graph on page 15 to answer these questions: *Which magnet pulled the paper clips the farthest? How many centimeters farther did the horseshoe magnet pull the paper clips than the bar magnet?*

- **OL** Have students read the book and answer Exercise 4. Have students use the graph on page 15 to answer these questions: *Which magnet pulled the nails the least distance? How many fewer centimeters did the bar magnet pull the nails than the horseshoe magnet?*

- **BL** Have students read the book and answer Exercises 4 and 5. Have students use the graph on page 15 to answer these questions: *How many centimeters farther did the horseshoe magnet pull the paper clips than the disk magnet? How many fewer centimeters did the disk magnet pull the nails than the horseshoe magnet?*

CCSS 2.MD.4   Use after Lesson 11-9

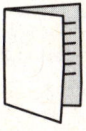

Real-World Problem Solving Reader *Fossils Over Time*

Summary *Fossils Over Time* answers many questions students have about fossils in general and dinosaurs in particular. Students will have experiences with measurements and tally charts.

Materials meterstick, paper, pencil

Note: Be sure students know how to use the meterstick to find their heights in centimeters.

- **AL** Have students read the book with a partner and answer Exercise 2.
- **OL** Have students read the book and answer Exercises 2 and 3.
- **BL** Have students read the book and answer Exercises 2 and 3. Have students look at pages 8 and 9 and find how many feet taller a *Brachiosaurus* is than a second grader. How many inches taller?

19

CHAPTER 11 **Customary and Metric Lengths** (continued)

 2.MD.4   Use after Lesson 11-10

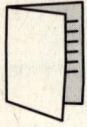

Real-World Problem Solving Reader *Animals Big and Small*

Summary In *Animals Big and Small*, students will have opportunities to compare sizes of animals, using nonstandard and standard units. Students will also compare adult animals to their offspring. This book contains a map to show area, charts, and Venn diagrams.

Materials paper, pencil

AL Have students read the book with a partner and answer Exercise 3. Have students look at the table on page 7 and compare the length of a baby polar bear with the length of an adult male polar bear. What is the difference in length?

OL Have students read the book and answer Exercise 3. Have students look at the table on page 7 and compare the length of a baby polar bear with the length of an adult male polar bear. What is the difference in length? Have them compare the length of an adult female polar bear and an adult male polar bear. What is the difference in length?

BL Have students read the book and answer Exercise 3. Have students look at the table on page 7 and compare the length of a baby polar bear with the length of an adult male polar bear. What is the difference in length? Have them compare the length of an adult female polar bear and an adult male polar bear. What is the difference in length? Have students look at page 8 and answer the following question: In 3 years, how many centimeters will the tusks of an adult elephant grow?

2.MD.3   Use after Lesson 11-7

Activity Card Centimeter Caterpillar

Materials drawing paper, crayons, centimeter ruler, classroom objects

AL Have students complete the Activity Card as written.

OL Have students complete the Activity Card as written, tracing and marking a line 20 centimeters long.

BL Have students complete the Activity Card as written, tracing and marking a line 30 centimeters long.

CHAPTER 12 Geometric Shapes and Equal Shares

The following learning station activities provide differentiated instructional strategies for helping students recognize two-dimensional and three-dimensional shapes and recognize equal shares of shapes.

 2.G.3   Use after Lesson 12-6

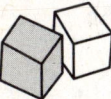

Game The Equalizer

Materials 32 two-color counters

Note: Before playing the game, discuss each shape on the board and how many equal shares are shaded.

  Have students play the game in teams of two.

  Have students play the game with the rules as written.

  Have students continue playing, but remove the counters as they find matches. Ask students to name each fraction as they remove the counters from the board.

 2.G.1   Use after Lesson 12-5

Graphic Novel Soccer Shape-Up

Materials blackline master pages 57 and 58, crayons, pencil

Watch the graphic novel as a class and discuss the following questions.

- What is happening? *Sample answer: The children are watching a soccer game and talking about the school project one of them has to complete.*
- What is the school project? *Sample answer: The boy needs to find three-dimensional shapes in the environment and take pictures of them.*

  Have students solve the graphic novel problem using blackline master page 57.

  Have students solve the graphic novel problem using blackline master page 57. On the back of the blackline master, have students draw another scene that includes three-dimensional shapes. Have students ask a friend to identify the three-dimensional shapes they see in the scene.

  Have students solve the graphic novel problem using blackline master page 57. Have students use blackline master page 58 to write their own graphic novels with a problem that requires them to identify three-dimensional shapes. Let students share their new graphic novels with a friend and ask the friend to solve the problem.

CHAPTER 12 **Geometric Shapes and Equal Shares** (continued)

 2.G.3   Use after Lesson 12-7

Graphic Novel The Case of the Missing Cake

Materials blackline master pages 59 and 60, crayons, pencil

Watch the graphic novel as a class and discuss the following questions.

- What is happening? *Sample answer: The children are playing basketball in the park. While they are playing, a dog eats some of their victory cake.*

- What will they do next? *Sample answer: They will find out how much of the cake is missing.*

AL   Have students solve the graphic novel problem using blackline master page 59.

OL   Have students solve the graphic novel problem using blackline master page 59. On the back of the blackline master, have students extend the graphic novel by drawing another cell to show what happens next.

BL   Have students solve the graphic novel problem using blackline master page 59. Have students use blackline master page 60 to write their own graphic novels with a problem that involves figuring out how much of the cake is missing and how much of the cake is left. Let students share their new graphic novels with a friend and ask the friend to solve the problem.

 2.G.1   Use after Lesson 12-5

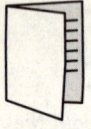

Real-World Problem Solving Reader *Homes of All Shapes*

Summary In *Homes of All Shapes*, students explore the variety in homes of people and animals. The book focuses on shapes, sizes, and functions of homes. Diagrams show differences in the sizes and shapes of animal and people homes.

Materials paper, pencil

   Have students read the book with a partner and answer Exercises 1 and 4.

   Have students read the book and answer Exercises 1, 2, 3, 4, and 6.

   Have students read the book and answer Exercises 1–6.

Answers

Answers and sample answers are provided for all the Graphic Novel, Real-World Problem Solving Reader, and Activity Card differentiated strategies.

CHAPTER 1 Apply Addition and Subtraction Concepts

> **Graphic Novel** *Puppies!*
>
> **AL** 11 − 5 = 6 puppies
>
> **OL** 11 − 5 = 6 puppies; Student graphic novel extensions and subtraction problems will vary. Accept all reasonable answers.
>
> **BL** 11 − 5 = 6 puppies; Student graphic novels and subtraction problems will vary. Accept all reasonable answers.

> **Real-World Problem Solving Reader** *How Many Seeds?*
>
> 1. 4 seeds
> 3. about 10 seeds
> 4. Sample answer: no; There aren't 10 seeds in the orange shown on the page.
> 5. Check students' drawings.
> 6. kiwi, pumpkin, orange, apple, peach

> **Activity Card** *Letter Subtraction*
>
> Student number sentences will vary. Accept all reasonable answers.

CHAPTER 2 **Number Patterns**

 Graphic Novel Flamingo Fun

 2 + 2 + 2 + 2 + 2 + 2 + 2 = 14, 14 legs in all

 2 + 2 + 2 + 2 + 2 + 2 + 2 = 14, 14 legs in all; Check student drawings. Accept all reasonable answers.

 2 + 2 + 2 + 2 + 2 + 2 + 2 = 14; 14 legs in all; Student graphic novels and skip-counting problems will vary. Accept all reasonable answers.

Real-World Problem Solving Reader Geese on the Go

1. 8 times
2. 15 geese
3. 10 eggs; 5 + 5 = 10
4. 20 hours
5. 12 geese; Sample answer: Two groups with six geese in each group is 12 (6 + 6 = 12).

Activity Card High-Frequency Words

Student answers will vary depending on the words chosen.

CHAPTER 3 **Add Two-Digit Numbers**

 Graphic Novel Tickets to the Puppet Show

 Sample answer: Yes, they met their goal (61 + 36 = 97; 97 is more than 90). They actually passed it.

 Sample answer: Yes, they met their goal (61 + 36 = 97; 97 is more than 90). They actually passed it. Student graphic novel extensions and two-digit addition problems will vary. Accept all reasonable answers.

 Sample answer: Yes, they met their goal (61 + 36 = 97; 97 is more than 90). They actually passed it. Student graphic novels and two-digit addition problems will vary. Accept all reasonable answers.

 Activity Card Telephone Words

Student responses will vary depending on the words they use. Accept all reasonable answers.

CHAPTER 4 Subtract Two-Digit Numbers

Graphic Novel *Swimming Fun*

- **AL** 20 − 9 = 11 (11 rings so far); 11 more times
- **OL** 20 − 9 = 11 (11 rings so far); 11 more times; Student problems will vary. Accept all reasonable answers.
- **BL** 20 − 9 = 11 (11 rings so far); 11 more times; Student graphic novels and two-digit subtraction problems will vary. Accept all reasonable answers.

Real-World Problem Solving Reader *Baseball's Hero*

1. 41 home runs and stolen bases
2. 3rd base; See students' explanations.
3. 51 bases
4. 1947 and 1948; The chart says he hit 12 home runs for both of those years.
5. 12 more bases

Activity Card Clay Sentences

Student responses will vary depending on the numbers students use. Accept all reasonable responses.

CHAPTER 5 Place Value to 1,000

Graphic Novel *Home Run Records*

- **AL** 203, 295, 516, 609; Andre's
- **OL** 203, 295, 516, 609; Andre's; Student graphic novel extensions and home run record numbers will vary. Accept all reasonable answers.
- **BL** 203, 295, 516, 609; Andre's; Student graphic novels and problems that require comparing three-digit numbers to solve will vary. Accept all reasonable answers.

Activity Card Comparing Stories

Student responses will vary depending on the numbers students choose to use in their tall tales.

CHAPTER 6 Add Three-Digit Numbers

Graphic Novel Enchanted Palace Park

- **AL** $395 + $395 = $790
- **OL** $395 + $395 = $790; $395 + $448 = $843
- **BL** $395 + $395 = $790; Student graphic novels and three-digit addition problems will vary. Accept all reasonable answers.

Real-World Problem Solving Reader Lady Liberty

- **AL** 700 pieces; 428 crates
- **OL** 700 pieces; 428 crates; 546 steps
- **BL** 700 pieces; 428 crates; $9.00; 546 steps

Real-World Problem Solving Reader Moving Along

- **AL** 660 miles; Sample answer: It is 500 miles from Hydro to Albuquerque. It is another 160 miles to Arizona. I added to find the total distance.
- **OL** 660 miles; Sample answer: It is 500 miles from Hydro to Albuquerque. It is another 160 miles to Arizona. I added to find the total distance.
  500 miles
- **BL** 660 miles; Sample answer: It is 500 miles from Hydro to Albuquerque. It is another 160 miles to Arizona. I added to find the total distance.
  500 miles; 202 years old

Activity Card How Many Calories?

Student responses will vary depending on the food labels they use. Accept all reasonable responses.

CHAPTER 7 Subtract Three-Digit Numbers

 Graphic Novel Ice Cream Favorites

- **AL** 191 − 109 = 82
- **OL** 191 − 109 = 82; Check student drawings. Accept all reasonable answers.
- **BL** 191 − 109 = 82; Student graphic novels and three-digit subtraction problems will vary. Accept all reasonable answers.

 Activity Card The Boiling Point

Student responses will vary depending on the numbers on the slips of paper they choose. Accept all reasonable answers.

- **BL** Student word problems will vary. Accept all reasonable responses.

CHAPTER 8 Money

 Graphic Novel Pinwheel, Please!

- **AL** 10¢, 20¢, 25¢ = 25¢; yes
- **OL** 10¢, 20¢, 25¢ = 25¢; yes; Student answers will vary. Accept all reasonable answers.
- **BL** 10¢, 20¢, 25¢ = 25¢; yes; Student graphic novels and money problems will vary. Accept all reasonable answers.

Real-World Problem Solving Reader *Our Grandma's Life*

1. 5 streets
2. Student answers will vary depending on the number of students in their class.
3. Student answers will vary depending on stamp price. Sample answer:
   42¢ + 42¢ = 84¢; 100¢ − 84¢ = 16¢; one dime, one nickel; one penny
4. 31; 31 + 40 = 71

 Activity Card Money Words

Student sentences and drawings will vary depending on the vocabulary words they have chosen. Accept all reasonable responses.

CHAPTER 9 Data Analysis

Graphic Novel And the Winner Is...

 AL   zoo

 OL   zoo; Student graphic novel extensions will vary. Accept all reasonable answers.

BL   zoo; Student graphic novels and problems that use information from a graph to solve will vary. Accept all reasonable answers.

Activity Card Say It With Pictures

Student responses will vary. Accept all reasonable answers.

CHAPTER 10 Time

Graphic Novel Pizza Time

 AL   6:00

 OL   6:00; Student graphic novel extensions and time problems will vary. Accept all reasonable answers.

 BL   6:00; Student graphic novels and time problems will vary. Accept all reasonable answers.

Real-World Problem Solving Reader A Mountain of Presidents

2. 11:00 A.M.
3. from 8:00 A.M. to 10:00 P.M.; 14 hours
4. 827 feet taller
BL   5:00 P.M.

Activity Card Time Journal

Student time journals will vary. Accept all reasonable responses.

CHAPTER 11 Customary and Metric Lengths

Graphic Novel Surfboard Selector

- AL 48 + 15 = 63, pink
- OL 48 + 15 = 63, pink; Check student drawings. Accept all reasonable answers.
- BL 48 + 15 = 63, pink; Student graphic novels and measurement problems will vary. Accept all reasonable answers.

Real-World Problem Solving Reader A Magnet's Strength

4. 52 nails
5. On page 11, the graphs show how many of the objects the magnets pulled. On page 15, the graphs show how far each of the objects are pulled; All the graphs show data about the same types of magnets; If stronger magnets had been used, the number of objects picked up and centimeters pulled would be larger.

- AL horseshoe magnet; 2 centimeters farther
- OL disk magnet; 2 fewer centimeters
- BL 2 centimeters farther; 3 fewer centimeters

Real-World Problem Solving Reader Fossils Over Time

2. Student answers will vary depending on their heights.
   Sample answer: I figured out my height in centimeters. Then I subtracted the dinosaur's length from my height.
3. 105 feet; greater because a *Stegosaurus* and *Tyrannosaurus* rex are 95 feet together.

- BL 46 feet taller; 552 inches taller

Real-World Problem Solving Reader Animals Big and Small

3. 61 centimeters

- AL 270 centimeters
- OL 270 centimeters; 50 centimeters
- BL 270 centimeters; 50 centimeters; 45 centimeters

Activity Card Centimeter Caterpillar

Student responses will vary depending on objects students are measuring. Accept all reasonable answers.

29

CHAPTER 12 **Geometric Shapes and Equal Shares**

Graphic Novel Soccer Shape-Up

AL Student responses will vary but may include the following: soccer ball—sphere; kite—rectangular prism; can—cylinder.

OL Student responses will vary but may include the following: soccer ball—sphere; kite—rectangular prism; can—cylinder. Student drawings will vary. Accept all reasonable answers.

BL Student responses will vary but may include the following: soccer ball—sphere; kite—rectangular prism; can—cylinder. Student graphic novels and three-dimensional shapes problems will vary. Accept all reasonable answers.

Graphic Novel The Case of the Missing Cake

AL Student drawings may vary but should show 4 equal parts with 1 part missing. 1 out of 4 parts is missing. One fourth of the cake is gone.

OL Student drawings may vary but should show 4 equal parts with 1 part missing. 1 out of 4 parts is missing. One fourth of the cake is gone. Student graphic novel extensions will vary. Accept all reasonable answers.

BL Student drawings may vary but should show 4 equal parts with 1 part missing. 1 out of 4 parts is missing. One fourth of the cake is gone. Student graphic novels and problems will vary. Accept all reasonable answers.

Real-World Problem Solving Reader Homes of All Shapes

1. Student answers will vary (students may count just large windows, just the smaller panes, or large windows and smaller panes).
2. See students' drawings
3. See students' drawings.
4. 6 faces
5. Sample answer: They are both three-dimensional shapes. Earth is a sphere and does not have sides and vertices. A pyramid does have sides and vertices.
6. See students' designs.

Game Board, Squares

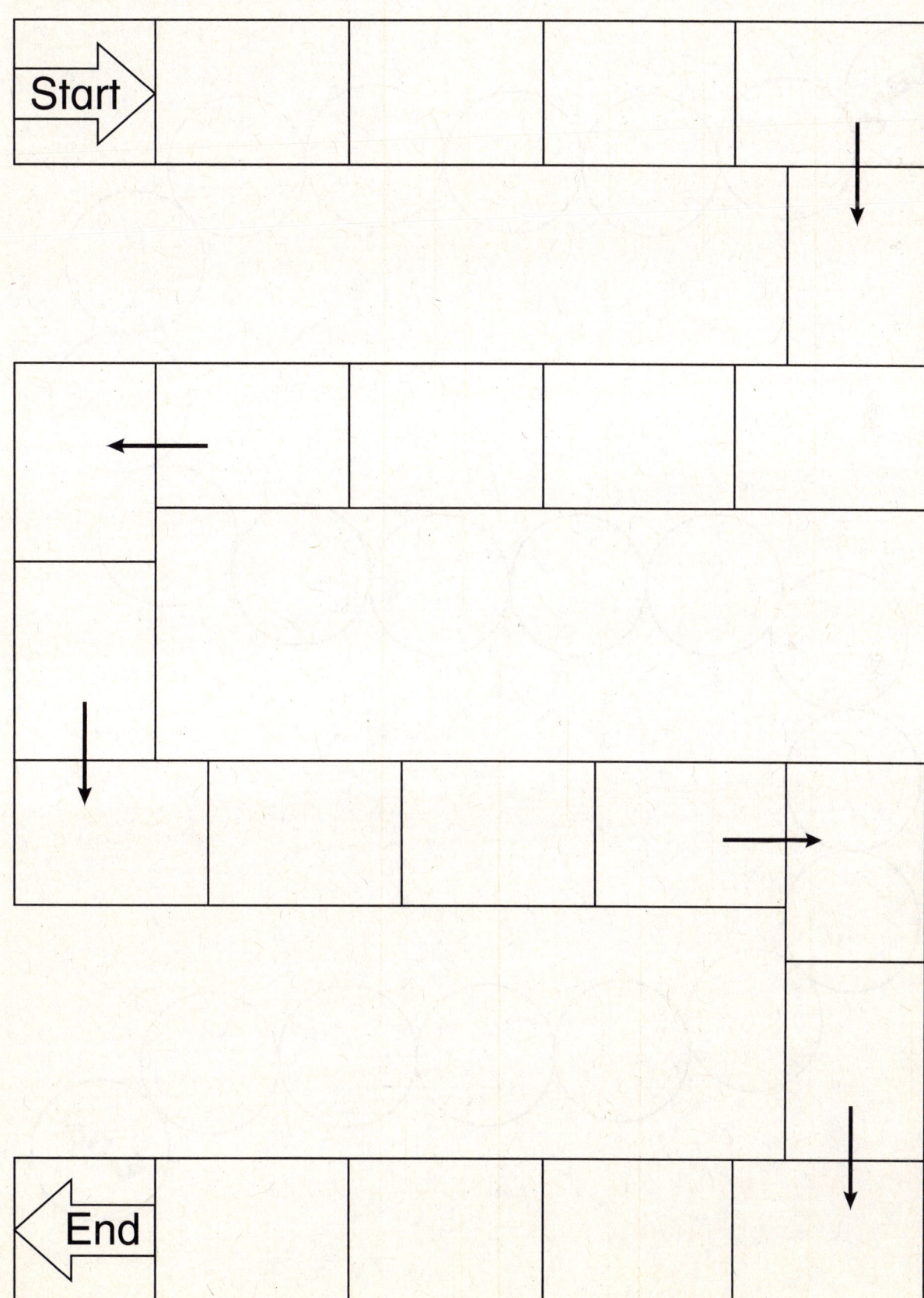

Name _____ Date _____

Game Board, Circles

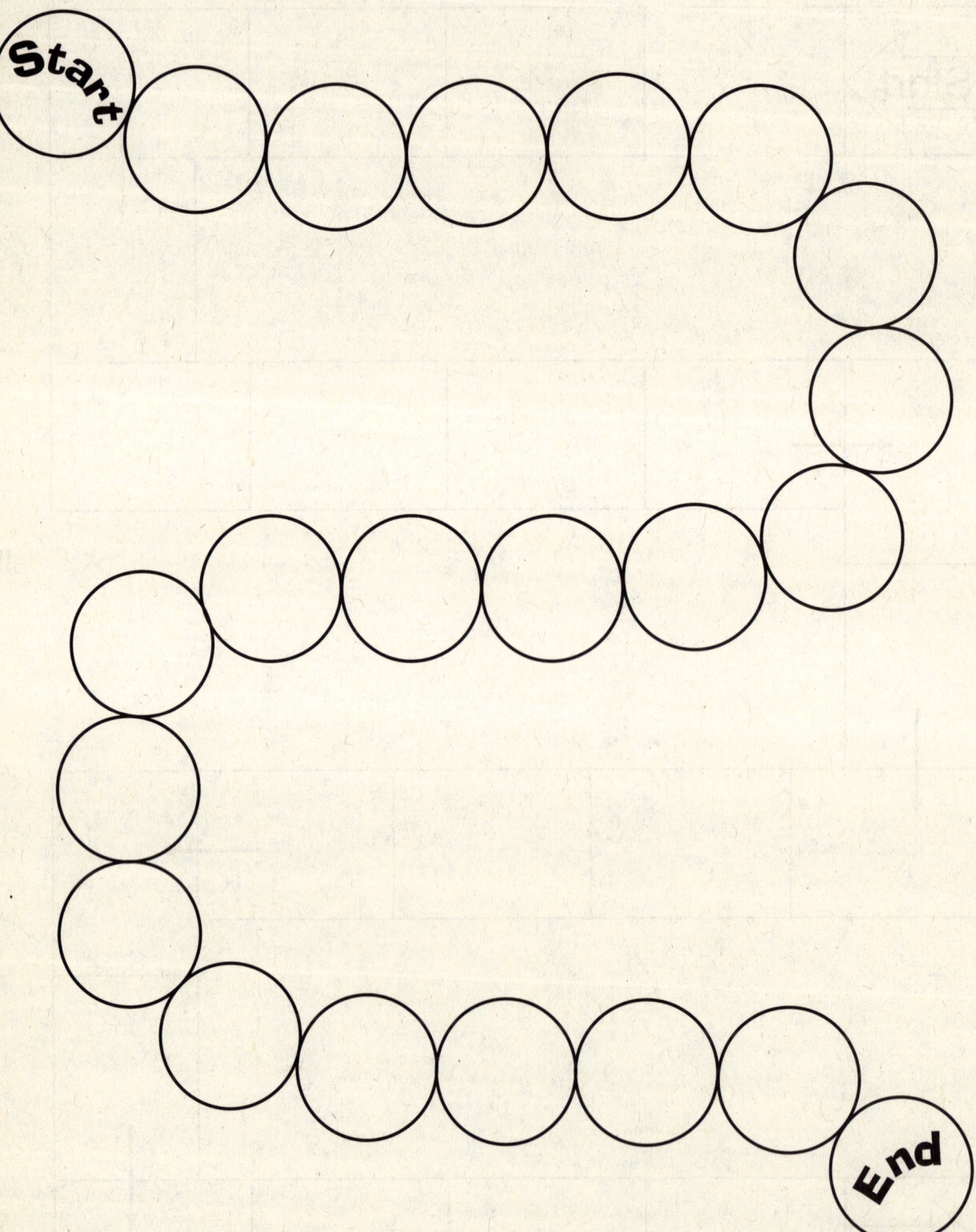

Name _____ Date _____

Use the information to solve the problem.

Flamingo Fun

Remember, we are trying to find a quicker way to count the flamingo legs.

I know! There are 7 flamingos. If we skip count by 2s, it will be much easier!

How many legs do 7 flamingos have?

____ + ____ + ____ + ____ + ____ + ____ + ____ = ____

____ legs in all

Name _____ Date _____

Kaya, Paul and Julie in Flamingo Fun

Name _____ Date _____

Use the information to solve the problems.

_____ + _____ = _____

Did they reach their goal? Explain how you know.

_____
_____

Name _____ Date _____

Use the information to solve the problems.

Swimming Fun

Remember, Julie is diving for rings. There are 20 rings in all.

How many rings do I have?

Let's count them. You have 9 rings.

How many rings so far? _____ − _____ = _____

If Julie brings up 1 ring
each time, she has to dive _____ more times.

39

Name _____ Date _____

Tens and Ones Chart

Tens	Ones

Name _____ Date _____

Use the information to solve the problems.

Home Run Records

Order from *least* to *greatest*.

_____ , _____ , _____ , _____

_____ player has the most home runs.

Name _____   Date _____

Use the information to solve the problem.

Enchanted Palace Park

How much are two passes? $ _____ + $ _____ = $ _____

Use the information given to solve the problem.

Ice Cream Favorites

_____ – _____ = _____ more students like chocolate.

Name _____   Date _____

Paul, Andre, Kaya, and Julie in
Ice Cream Favorites

Name _____ Date _____

Use the information to solve the problems.

Pinwheel, Please!

Remember, tomorrow is my little sister's birthday, and I want to get her a pinwheel.

Sophia has two dimes and a nickel.

The sign says they cost 25¢ each. Does she have enough?

25¢ each

_____ ¢, _____ ¢, _____ ¢ = _____ ¢   Do I have enough? _____

Use the information to solve the problem.

And the Winner Is...

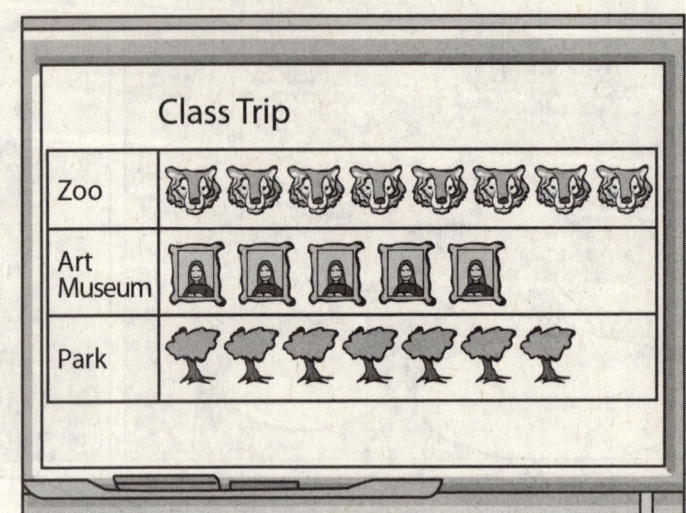

Which place received the most votes? _____

Name _____ Date _____

Analog Clockface

1. Mount on heavy paper.
2. Cut out the clock hands.
3. Attach them to the clock with a paper fastener.

Use the information to solve the problem.

Pizza Time

What time does gymnastics class start? _____

Name _____ Date _____

Use the information to solve the problems.

Surfboard Selector

_____ + _____ = _____ inches

Kaya has to pick the _____ board.

Name _____ Date _____

Paul, Julie and Kaya in Surfboard Selector

Use the information to solve the problem.

Soccer Shape-Up

Which three-dimensional figures do you see?
_____
_____

Name _____ Date _____

Kaya, Omar and Paul in
Soccer Shape-up

Name _____ Date _____

Use the information to solve the problems.

The Case of the Missing Cake

Draw what is left of the cake. Divide it into equal parts. Number each part. _____ out of _____ parts is missing.

_____ of the cake is gone.